FRACTIONS
The Meaning of Fractions

Allan D. Suter

McGraw Hill Wright Group

Series Editor: Mitch Rosin
Executive Editor: Linda Kwil
Production Manager: Genevieve Kelley
Marketing Manager: Sean Klunder
Cover Design: Steve Strauss, ¡Think! Design

Wright Group

Send all inquiries to:
McGraw-Hill/Contemporary
130 East Randolph Street, Suite 400
Chicago, Illinois 60601

ISBN: 0-07-287108-3

Printed in the United States of America.

2 3 4 5 6 7 8 9 10 QPD/QPD 09 08 07 06 05

The **McGraw·Hill** Companies

◼ Contents

1. Which two fractions are represented in this drawing:

 $\frac{2}{4}$, $\frac{1}{3}$, or $\frac{1}{2}$?

Answer: _____

2. If a whole is divided into 6 equal parts, 1 part is what fraction of the whole?

Answer: _____

3. Which is larger, $\frac{1}{7}$ or $\frac{1}{8}$?

Answer: _____

4. What is largest, $\frac{1}{6}$, $\frac{3}{6}$, or $\frac{5}{6}$?

Answer: _____

5. Change $\frac{1}{2}$ to an equivalent fraction with a denominator of 12.

Answer: _____

6. Change $\frac{2}{3}$ to an equivalent fraction with a denominator of 15.

Answer: _____

7. Change $\frac{8}{9}$ to an equivalent fraction with a denominator of 18.

Answer: _____

8. Change $\frac{3}{4}$ to an equivalent fraction with a denominator of 28.

Answer: _____

9. On a quiz with 10 questions, Sam got 3 wrong. What fraction of the questions did Sam get wrong?

Answer: _____

10. Jed has 12 pairs of socks. 8 pairs of his socks are white. What fraction of his socks are white? Express the answer in simplest form.

Answer: _____

11. A pizza was cut into 8 equal slices. Tim and Sharon ate 6 slices. What fraction of the pizza did they eat? Express the answer in simplest form.

Answer: _____

12. A yard is equal to 36 inches. 16 inches are what fraction of a yard? Express the answer in simplest form.

Answer: _____

13. Change $\frac{8}{32}$ to a fraction in simplest form.

Answer: _____

14. Write $\frac{4}{12}$ in simplest form.

Answer: _____

15. Write $\frac{6}{10}$ in simplest form.

Answer: _____

16. Change $\frac{35}{50}$ to a fraction in simplest form.

Answer: _____

17. What is the decimal value of $\frac{3}{8}$ to three decimal places?

Answer: _____

18. Write $\frac{1}{6}$ in decimal form to the hundredths place. Write any remainder as a fraction.

Answer: _____

19. Find the decimal form of $\frac{7}{4}$.

Answer: _____

20. What is the decimal value of $\frac{11}{10}$?

Answer: _____

Evaluation Chart

On the following chart, circle the number of any problem you missed. The column after the problem number tells you the pages where those problems are taught. Based on your score, your teacher may ask you to study specific sections of this book. However, to thoroughly review your skills, begin with Unit 1 on page 7.

Skill Area	Pretest Problem Number	Skill Section	Review Page
Identifying Fractions	1, 2	7–19	20
Comparing Fractions	3, 4	21–25	26
Equivalent Fractions	5, 6, 7, 8	27–33	34
Comparisons	9, 10, 11, 12	35–39	40
Simplifying Fractions	13, 14, 15, 16	41–57	58
Fractions and Decimals	17, 18, 19, 20	59–62	63
Life-Skills Math	All	64–73	72, 74

Fractions: The Meaning of Fractions

Denominators

<table>
<tr><td>The bottom number of a fraction—the denominator—tells how many equal parts a whole object is divided into.</td><td>$\dfrac{1}{4}$ ← denominator</td></tr>
</table>

Below are drawings of whole objects that are divided into equal parts. Write the fraction that is shaded in each drawing.

 1.   **2.**   **3.**   **4.**   **5.**   **6.**

 shaded part → $\dfrac{1}{4}$
total equal parts → ___ ___ ___ ___ ___

7.   **8.**   **9.**   **10.**   **11.**   **12.**

denominator → $\dfrac{1}{8}$ ___   ___   ___   ___   ___

Shade **only one** of the equal parts in each drawing. Write a fraction for the shaded part of each object.

13.   **14.**   **15.**   **16.**   **17.**

shaded part → $\dfrac{1}{5}$
total equal parts → ___ ___ ___ ___

18.   **19.**   **20.**   **21.**   **22.**

      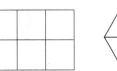

$\dfrac{1}{4}$ ___   ___   ___   ___

Numerators

The top number of the fraction—the **numerator**—
tells how many equal parts are used in the whole object.

$\dfrac{1}{4}$ ◀— numerator

Below are drawings of whole objects that are divided into equal parts. Write
the fraction that is shaded in each drawing.

1.

fraction ⟶ $\dfrac{3}{4}$
shaded

2.

___

3.

___

4.

___

5.

numerator ⟶ $\dfrac{2}{4}$

6.

___

7.

___

8.

___

9.

___

10.

___

11.

___

12.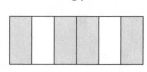

___

13.

___

14.

___

15.

___

Fractions with Numerators of 1

Match each drawing with the correct fraction.

1. $\left(\dfrac{1}{5}\right)$ 2. $\left(\dfrac{1}{8}\right)$ 3. $\left(\dfrac{1}{9}\right)$

$\underline{\quad D \quad}$ $\underline{\qquad}$ $\underline{\qquad}$
letter   letter   letter

A   **B**

4. $\left(\dfrac{1}{7}\right)$ 5. $\left(\dfrac{1}{10}\right)$ 6. $\left(\dfrac{1}{3}\right)$

$\underline{\qquad}$ $\underline{\qquad}$ $\underline{\qquad}$
letter   letter   letter

C   **D**   **E**

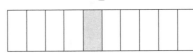

F   **G**   **H**

7. $\left(\dfrac{1}{4}\right)$ 8. $\left(\dfrac{1}{6}\right)$ 9. $\left(\dfrac{1}{2}\right)$

$\underline{\qquad}$ $\underline{\qquad}$ $\underline{\qquad}$
letter   letter   letter

I

Draw and shade in each of the fractions below using the rectangles.

10. $\dfrac{1}{5}$ 11. $\dfrac{1}{6}$ 12. $\dfrac{1}{3}$ 13. $\dfrac{1}{7}$

shade in

14. $\dfrac{1}{2}$ 15. $\dfrac{1}{8}$ 16. $\dfrac{1}{4}$ 17. $\dfrac{1}{9}$

Match the Fractions

A fraction with a numerator smaller than its denominator is a **proper fraction.**

Match each drawing with the correct proper fraction.

1. $\left(\dfrac{2}{3}\right)$ 2. $\left(\dfrac{3}{6}\right)$

$\dfrac{A}{\text{letter}}$ $\dfrac{}{\text{letter}}$

3. $\left(\dfrac{1}{4}\right)$ 4. $\left(\dfrac{5}{6}\right)$

$\dfrac{}{\text{letter}}$ $\dfrac{}{\text{letter}}$

5. $\left(\dfrac{3}{5}\right)$ 6. $\left(\dfrac{2}{4}\right)$

$\dfrac{}{\text{letter}}$ $\dfrac{}{\text{letter}}$

7. $\left(\dfrac{5}{8}\right)$ 8. $\left(\dfrac{3}{4}\right)$

$\dfrac{}{\text{letter}}$ $\dfrac{}{\text{letter}}$

9. $\left(\dfrac{1}{3}\right)$ 10. $\left(\dfrac{1}{2}\right)$

$\dfrac{}{\text{letter}}$ $\dfrac{}{\text{letter}}$

A   **B**   **C**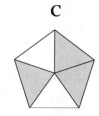

D   **E**   **F**

G   **H**

I **J**

Draw and shade in each of the fractions below using the rectangles.

11. $\dfrac{2}{5}$ 12. $\dfrac{3}{4}$ 13. $\dfrac{1}{6}$ 14. $\dfrac{3}{5}$

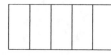
shade in

15. $\dfrac{2}{3}$ 16. $\dfrac{3}{8}$ 17. $\dfrac{1}{2}$ 18. $\dfrac{3}{7}$

Fractions Larger Than 1

A fraction with a numerator larger than its denominator is an **improper fraction**.

Count the shaded parts and complete the number sentence for each drawing.

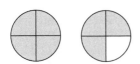

1. $\dfrac{4}{4}$ + $\dfrac{3}{4}$ = $\dfrac{7}{4}$
    _____
    answer

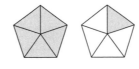

2. $\dfrac{5}{5}$ + $\dfrac{1}{5}$ = _____
    answer

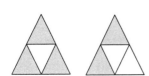

3. $\dfrac{2}{2}$ + $\dfrac{2}{2}$ + $\dfrac{1}{2}$ = _____
    answer

4. $\dfrac{1}{2}$ + $\dfrac{1}{2}$ + $\dfrac{2}{2}$ = _____
    answer

5. $\dfrac{3}{4}$ + $\dfrac{2}{4}$ = _____
    answer

6. $\dfrac{3}{3}$ + $\dfrac{3}{3}$ + $\dfrac{1}{3}$ = _____
    answer

Draw and shade in each of the improper fractions below using the rectangles.

7. $\dfrac{3}{2}$

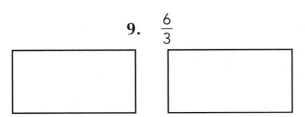

shade in

8. $\dfrac{5}{4}$

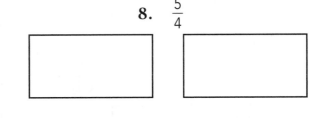

9. $\dfrac{6}{3}$

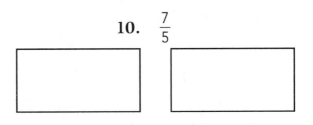

10. $\dfrac{7}{5}$

Mixed Numbers

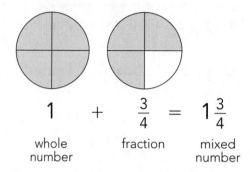

$$1 \quad + \quad \frac{3}{4} \quad = \quad 1\frac{3}{4}$$

whole   fraction   mixed
number   number

A **mixed number** combines a whole number and a proper fraction.

Write the mixed numbers.

1.

$$1 \quad + \quad 1 \quad + \quad \frac{1}{2} \quad = \quad 2\frac{1}{2}$$

mixed number

2.

$$1 \quad + \quad \frac{2}{3} \quad = \quad \underline{\hspace{2cm}}$$

mixed number

3.

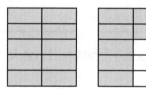

$$1 \quad + \quad \frac{7}{10} \quad = \quad \underline{\hspace{2cm}}$$

mixed number

4.

$$1 \quad + \quad 1 \quad + \quad \frac{3}{4} = \quad \underline{\hspace{2cm}}$$

mixed number

5. $2 + \frac{5}{7} =$ _____
mixed number

6. $5 + \frac{2}{9} =$ _____
mixed number

7. $1 + \frac{7}{8} =$ _____
mixed number

8. $7 + \frac{1}{2} =$ _____
mixed number

9. $3 + \frac{2}{3} =$ _____
mixed number

10. $6 + \frac{3}{5} =$ _____
mixed number

11. $2 + \frac{1}{6} =$ _____
mixed number

12. $1 + \frac{1}{4} =$ _____
mixed number

Match the Mixed Numbers

Remember: a mixed number combines a whole number and a proper fraction.

Examples: $2 + \dfrac{2}{3} = 2\dfrac{2}{3}$ $9 + \dfrac{3}{4} = 9\dfrac{3}{4}$

Match each drawing with the correct mixed number.

1. $\left(2\dfrac{2}{3}\right)$ 2. $\left(1\dfrac{3}{4}\right)$

  C
  ‾‾‾ ‾‾‾
  letter letter

A   B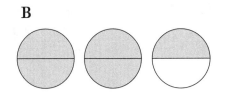

3. $\left(1\dfrac{1}{4}\right)$ 4. $\left(1\dfrac{2}{6}\right)$

  ‾‾‾ ‾‾‾
  letter letter

C 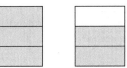   D

5. $\left(1\dfrac{2}{3}\right)$ 6. $\left(2\dfrac{1}{4}\right)$

  ‾‾‾ ‾‾‾
  letter letter

E    F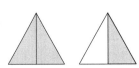

7. $\left(2\dfrac{1}{2}\right)$ 8. $\left(1\dfrac{1}{2}\right)$

  ‾‾‾ ‾‾‾
  letter letter

G 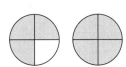   H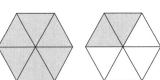

Draw rectangles and shade in each of the mixed numbers below.

9. $1\dfrac{1}{4}$—One and one fourth

10. $2\dfrac{2}{5}$—Two and two fifths

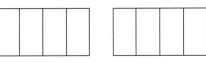

shade in

11. $2\dfrac{1}{3}$—Two and one third

12. $2\dfrac{1}{5}$—Two and one fifth

Writing Fractions and Mixed Numbers

For each of the shaded figures, write the improper fraction and the mixed number.

Improper Fraction		Mixed Number	

1. $\dfrac{3}{2}$ = $1\dfrac{1}{2}$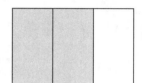

2. $\dfrac{}{3}$ = $\dfrac{}{}$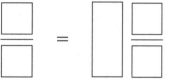

3. $\dfrac{}{}$ = $\dfrac{}{}$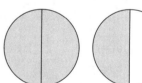

4. $\dfrac{}{}$ = $\dfrac{}{}$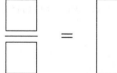

5. $\dfrac{}{}$ = $\dfrac{}{}$

6. $\dfrac{}{}$ = $\dfrac{}{}$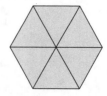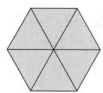

Fractions: The Meaning of Fractions

Shade the Mixed Numbers

Shade in the figures to show the mixed numbers.

1. $1\frac{1}{2}$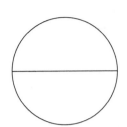

2. $1\frac{3}{4}$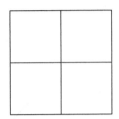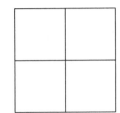

3. $1\frac{3}{5}$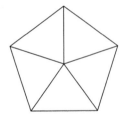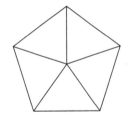

4. $2\frac{3}{8}$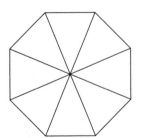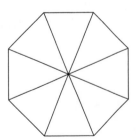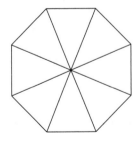

5. $2\frac{5}{6}$

Write the Numbers

Write the fractions, mixed numbers, or whole numbers shown below.

1. If $\boxed{\begin{array}{ccc} \frac{1}{6} & \frac{1}{6} & \frac{1}{6} \\ \frac{1}{6} & \frac{1}{6} & \frac{1}{6} \end{array}}$ = 1, then 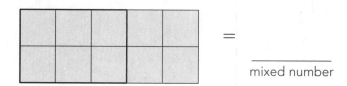 = $1\frac{2}{6}$

   _____
   mixed number

2. If [grid] = 1, then [grid] =

   _____
   mixed number

3. If [grid] = 1, then [grid] =

   _____
   fraction

4. If [grid] = 1, then [grid] =

   _____
   mixed number

5. If $\boxed{\frac{1}{4}}\ \boxed{\frac{1}{4}}\ \boxed{\frac{1}{4}}\ \boxed{\frac{1}{4}}$ = 1, then $\boxed{\frac{1}{4}}\ \boxed{\frac{1}{4}}\ \boxed{\frac{1}{4}}\ \boxed{\frac{1}{4}}\ \boxed{\frac{1}{4}}$ = $1\frac{1}{4}$

   _____
   mixed number

6. If [squares] = 1, then [squares] =

   _____
   whole number

7. If [squares] = 1, then [squares] =

   _____
   fraction

8. If [squares] = 1, then [squares] =

   _____
   mixed number

Sets

Match each drawing with the correct answer.

1. $\frac{1}{2}$ of 8

   H
   letter

2. $\frac{1}{2}$ of 4

   _____
   letter

3. $\frac{1}{3}$ of 3

   _____
   letter

4. $\frac{1}{2}$ of 6

   _____
   letter

5. $\frac{1}{2}$ of 2

   _____
   letter

6. $\frac{1}{3}$ of 6

   _____
   letter

7. $\frac{1}{4}$ of 8

   _____
   letter

8. $\frac{1}{2}$ of 3

   _____
   letter

A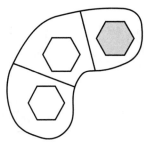

B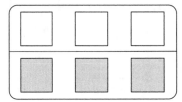

C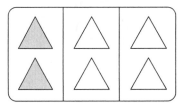

D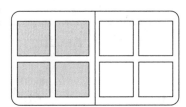

E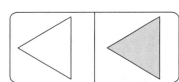

F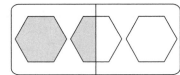

G

H

Shade the drawings.

9. $\frac{1}{2}$ of 4

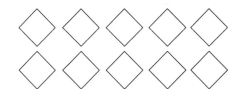

10. $\frac{1}{3}$ of 6

11. $\frac{1}{2}$ of 10

12. $\frac{1}{4}$ of 8

More Sets

Match each drawing with the correct answer.

1. 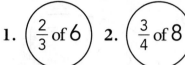 $\frac{2}{3}$ of 6   2. $\frac{3}{4}$ of 8

     E

   letter   letter

3. $\frac{4}{5}$ of 5   4. $\frac{3}{5}$ of 10

   letter   letter

5. $\frac{2}{3}$ of 3

   letter

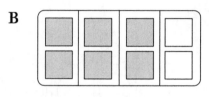

A

B

C

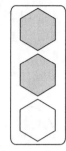

D

E

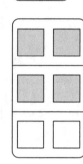

Shade the drawings.

6. $\frac{3}{4}$ of 4

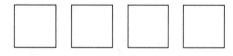

7. $\frac{2}{3}$ of 6

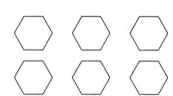

8. $\frac{4}{5}$ of 10

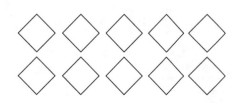

9. $\frac{3}{4}$ of 12

Looking at Sets

1. Shade $\frac{1}{2}$ of the six squares.

2. Circle $\frac{2}{3}$ of the nine Xs.

3. Shade $\frac{1}{4}$ of the twelve squares.

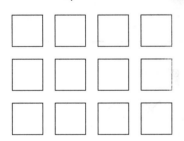

4. Circle $\frac{1}{3}$ of the twelve Xs.

5. Shade $\frac{3}{5}$ of the ten squares.

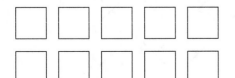

6. Shade $\frac{4}{5}$ of the ten squares.

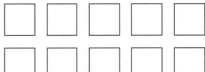

7. Circle $\frac{3}{4}$ of the twelve Xs.

8. Shade $\frac{1}{2}$ of the eight squares.

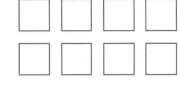

9. Circle $\frac{5}{6}$ of the six Xs.

10. Shade $\frac{2}{3}$ of the eighteen squares.

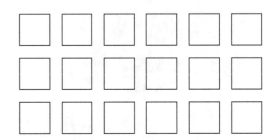

Identifying Fractions Review

Answer each problem.

1. Shade $1\frac{1}{4}$.

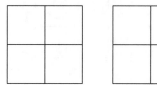

5. Shade $\frac{4}{5}$ of the ten squares.

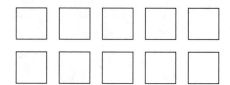

2. Write the fraction.

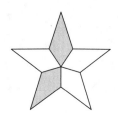

Answer: ——

6. Shade $\frac{1}{3}$ of the 12 squares.

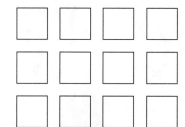

3. Add the shaded sections and write the answer as a fraction.

Answer: ——

7. Circle $\frac{5}{6}$ of the 12 Xs.

4. Write a fraction to show the number of crushed cans.

Answer: ——

8. Circle $\frac{1}{4}$ of the 16 Xs.

Fractions: The Meaning of Fractions

2 ■ Comparing Fractions

Comparing Fractions with the Same Denominator

Shade

$\frac{2}{6}$

Shade

$\frac{5}{6}$

Shade

$\frac{1}{6}$

Shade

$\frac{3}{6}$

Shade

$\frac{6}{6}$

Shade

$\frac{4}{6}$

Arrange the fractions above from smallest to largest.

7. _____ _____ _____ _____ _____ _____
   smallest   largest

8. Shade
   $\frac{1}{4}$

9. Shade
   $\frac{3}{4}$

10. Shade
    $\frac{2}{4}$

11. Shade
    $\frac{0}{4}$

12. Shade
    $\frac{4}{4}$

Arrange the fractions above from smallest to largest.

13. _____ _____ _____ _____ _____
    smallest   largest

If two fractions have the same denominator:

14. The fraction with the larger numerator is _____.
                                                smaller or larger

15. The fraction with the smaller numerator is _____.
                                                 smaller or larger

Comparing Fractions with the Same Numerator

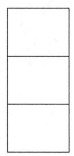 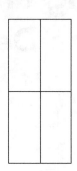 

1. Shade
$\dfrac{2}{3}$

2. Shade
$\dfrac{2}{12}$

3. Shade
$\dfrac{2}{4}$

4. Shade
$\dfrac{2}{8}$

Arrange the fractions above from smallest to largest.

5. _____ _____ _____ _____
smallest   largest

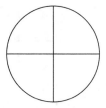 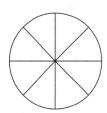  

6. Shade
$\dfrac{3}{4}$

7. Shade
$\dfrac{3}{8}$

8. Shade
$\dfrac{3}{5}$

9. Shade
$\dfrac{3}{3}$

Arrange the fractions above from smallest to largest.

10. _____ _____ _____ _____
smallest   largest

If two fractions have the same numerator:

11. The fraction with the larger denominator is _____.
smaller or larger

12. The fraction with the smaller denominator is _____.
smaller or larger

Shade and Compare

1. $\frac{2}{5}$

2. $\frac{1}{2}$

3. Larger fraction _____

4. Smaller fraction _____

5. Shade $\frac{5}{6}$

6. Shade $\frac{1}{2}$

7. Larger fraction _____

8. Smaller fraction _____

9. Shade $\frac{4}{5}$

10. Shade $\frac{1}{2}$

11. Larger fraction _____

12. Smaller fraction _____

13. Shade $\frac{1}{3}$

14. Shade $\frac{1}{2}$

15. Larger fraction _____

16. Smaller fraction _____

17. Shade $\frac{5}{6}$

18. Shade $\frac{4}{6}$

19. Larger fraction _____

20. Smaller fraction _____

21. Shade $\frac{3}{4}$

22. Shade $\frac{6}{8}$

23. What do you notice about these two fractions? _____

24. Shade $\frac{5}{3}$

25. Shade $\frac{3}{2}$

26. Larger fraction _____

27. Smaller fraction _____

28. Shade $\frac{2}{3}$

29. Shade $\frac{5}{6}$

30. Larger fraction _____

31. Smaller fraction _____

Order the Fractions

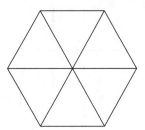 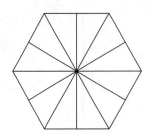 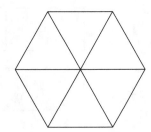

1. Shade $\frac{5}{6}$ **2.** Shade $\frac{11}{12}$ **3.** Shade $\frac{1}{6}$

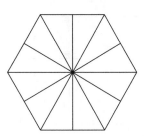 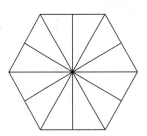 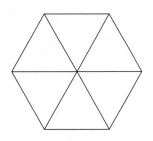

4. Shade $\frac{1}{12}$ **5.** Shade $\frac{5}{12}$ **6.** Shade $\frac{3}{6}$

Compare the size of the shaded parts in each drawing. Arrange the fractions above in order from smallest to largest.

7. _____ _____ _____ _____ _____ _____
    smallest   largest

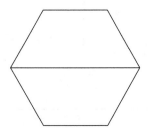 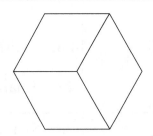 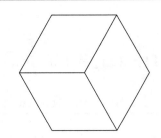

8. Shade $\frac{1}{2}$ **9.** Shade $\frac{2}{3}$ **10.** Shade $\frac{1}{3}$

Compare the size of the shaded parts in each drawing. Arrange the fractions above in order from smallest to largest.

11. _____ _____ _____
    smallest   largest

Practice Helps

Complete the problems.

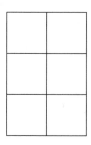

1. Shade $\frac{1}{3}$

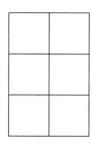

2. Shade $\frac{1}{2}$

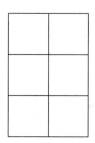

3. Shade $\frac{5}{6}$

Compare the size of the shaded parts in each drawing. Arrange the fractions above in order from largest to smallest.

4. _____ _____ _____
   smallest   largest

5.

Shade $\frac{1}{3}$

6.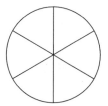

Shade $\frac{5}{6}$

7.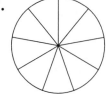

Shade $\frac{2}{9}$

Compare the size of the shaded parts in each drawing. Arrange the fractions above in order from smallest to largest.

8. _____ _____ _____
   smallest   largest

Comparing Fractions Review

Answer each problem.

1. Shade $\frac{4}{7}$

2. Shade $\frac{5}{7}$

3. Larger fraction _____

4. Smaller fraction _____

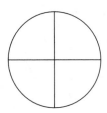 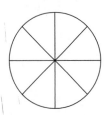  

5. Shade $\frac{1}{4}$ 6. Shade $\frac{3}{8}$ 7. Shade $\frac{3}{5}$ 8. Shade $\frac{3}{3}$

9. Arrange the fractions from smallest to largest.

_____ _____ _____ _____
smallest   largest

10. Shade $\frac{4}{3}$

11. Shade $\frac{3}{2}$

12. Larger fraction _____

13. Smaller fraction _____

What Are Equivalent Fractions?

A   B

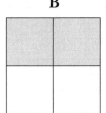

1. Are the two squares A and B the same size? _____

2. What fraction is shaded in square A? _____

3. What fraction is shaded in square B? _____

4. Are the unshaded parts in both A and B the same size? _____

5. Are the shaded parts in A and B the same size? _____

6. Does $\frac{1}{2} = \frac{2}{4}$? _____

Fractions that have the same (equal) value are **equivalent fractions.**

Look at the shaded parts in the figures below and fill in the equivalent fractions.

7.

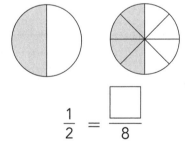

$$\frac{1}{2} = \frac{\boxed{}}{8}$$

8.

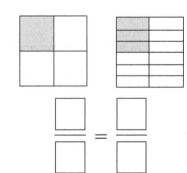

$$\frac{\boxed{}}{\boxed{}} = \frac{\boxed{}}{\boxed{}}$$

9.

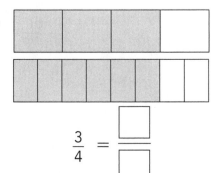

$$\frac{3}{4} = \frac{\boxed{}}{\boxed{}}$$

10.

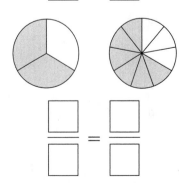

$$\frac{\boxed{}}{\boxed{}} = \frac{\boxed{}}{\boxed{}}$$

Shaded Equivalent Fractions

1. Shade $\frac{1}{2}$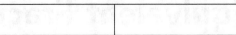

2. Shade $\frac{2}{4}$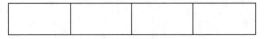

3. Shade $\frac{4}{8}$

4. Are the unshaded parts the same size in each of the rectangles? _____

5. Are the shaded parts in each rectangle the same size? _____

6. Complete: $\dfrac{1}{2} = \dfrac{}{4} = \dfrac{}{8}$

7.

   Shade $\frac{1}{3}$

8.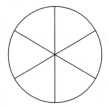

   Shade $\frac{2}{6}$

9.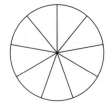

   Shade $\frac{3}{9}$

10. Complete: $\dfrac{1}{3} = \dfrac{}{6} = \dfrac{}{9}$

11.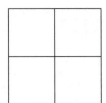

    Shade $\frac{3}{4}$

12.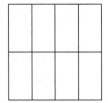

    Shade $\frac{6}{8}$

13.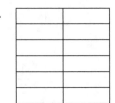

    Shade $\frac{9}{12}$

14. Complete: $\dfrac{3}{4} = \dfrac{}{8} = \dfrac{}{12}$

Writing Equivalent Fractions

1															
$\frac{1}{2}$								$\frac{1}{2}$							
$\frac{1}{4}$				$\frac{1}{4}$				$\frac{1}{4}$				$\frac{1}{4}$			
$\frac{1}{8}$		$\frac{1}{8}$		$\frac{1}{8}$		$\frac{1}{8}$		$\frac{1}{8}$		$\frac{1}{8}$		$\frac{1}{8}$		$\frac{1}{8}$	
$\frac{1}{16}$	$\frac{1}{16}$	$\frac{1}{16}$	$\frac{1}{16}$	$\frac{1}{16}$	$\frac{1}{16}$	$\frac{1}{16}$	$\frac{1}{16}$	$\frac{1}{16}$	$\frac{1}{16}$	$\frac{1}{16}$	$\frac{1}{16}$	$\frac{1}{16}$	$\frac{1}{16}$	$\frac{1}{16}$	$\frac{1}{16}$

EXAMPLE 1

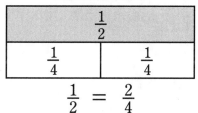

$$\frac{1}{2} = \frac{2}{4}$$

EXAMPLE 2

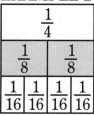

$$\frac{1}{4} = \frac{2}{8} = \frac{4}{16}$$

Use the chart to fill in the equivalent fractions below.

1. $\frac{1}{2} = \frac{}{4}$

2. $\frac{1}{4} = \frac{}{8}$

3. $\frac{1}{8} = \frac{}{16}$

4. $\frac{1}{2} = \frac{}{8}$

5. $\frac{3}{4} = \frac{}{8}$

6. $\frac{4}{8} = \frac{}{4}$

7. $\frac{2}{16} = \frac{}{8}$

8. $\frac{3}{8} = \frac{}{16}$

9. $1 = \frac{}{8}$

10. $\frac{6}{16} = \frac{}{8}$

11. $\frac{5}{8} = \frac{}{16}$

12. $\frac{4}{16} = \frac{}{4}$

13. $\frac{10}{16} = \frac{}{8}$

14. $1 = \frac{}{2}$

15. $\frac{3}{4} = \frac{}{16}$

16. $\frac{8}{16} = \frac{}{2}$

17. $\frac{3}{4} = \frac{}{8} = \frac{}{16}$

18. $\frac{1}{2} = \frac{}{4} = \frac{}{16}$

19. $1 = \frac{}{2} = \frac{}{8}$

20. $\frac{4}{16} = \frac{}{8} = \frac{}{4}$

21. $\frac{16}{16} = \frac{}{8} = \frac{}{4}$

22. $\frac{4}{4} = \frac{}{2} = \frac{}{1}$

23. $\frac{2}{8} = \frac{}{4} = \frac{}{16}$

24. $\frac{12}{16} = \frac{}{4} = \frac{}{8}$

More Equivalent Fractions

You cannot use charts all the time to compare fractions. To find equivalent fractions, multiply both the numerator and denominator by the same number.

A fraction with the same numerator and denominator is equal to 1.

$$\frac{2}{5} = \frac{2}{5} \times \boxed{\frac{2}{2}} = \frac{2 \times 2}{5 \times 2} = \frac{4}{10}$$

Different form, same value.

1. $\dfrac{1}{3} = \dfrac{1 \times \boxed{5}}{3 \times \boxed{5}} = \dfrac{\boxed{}}{15}$

2. $\dfrac{3}{4} = \dfrac{3 \times \boxed{3}}{4 \times \boxed{3}} = \dfrac{\boxed{}}{12}$

3. $\dfrac{5}{8} = \dfrac{5 \times \boxed{4}}{8 \times \boxed{4}} = \dfrac{\boxed{}}{32}$

4. $\dfrac{2}{3} = \dfrac{2 \times 4}{3 \times 4} = \dfrac{\boxed{}}{12}$

5. $\dfrac{1}{2} = \dfrac{1 \times 7}{2 \times 7} = \dfrac{\boxed{}}{14}$

6. $\dfrac{3}{5} = \dfrac{3 \times 4}{5 \times 4} = \dfrac{\boxed{}}{20}$

7. $\dfrac{3}{7} = \dfrac{3 \times 5}{7 \times 5} = \dfrac{\boxed{}}{35}$

8. $\dfrac{5}{12} = \dfrac{5 \times 2}{12 \times 2} = \dfrac{\boxed{}}{24}$

9. $\dfrac{5}{6} = \dfrac{5 \times 3}{6 \times 3} = \dfrac{\boxed{}}{18}$

10. $\dfrac{7}{9} = \dfrac{7 \times 7}{9 \times 7} = \dfrac{\boxed{}}{63}$

11. $\dfrac{3}{8} = \dfrac{3 \times 7}{8 \times 7} = \dfrac{\boxed{}}{56}$

12. $\dfrac{7}{12} = \dfrac{7 \times 4}{12 \times 4} = \dfrac{\boxed{}}{48}$

Using Fractions Equal to 1

Multiplying a number or fraction by 1 does not change its value.

To find an equivalent fraction:

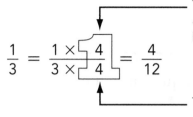

Think: what number can you multiply 1 by to equal 4?

$$\frac{1}{3} = \frac{1 \times 4}{3 \times 4} = \frac{4}{12}$$

Think: what number can you multiply 3 by to equal 12?

so $\frac{1}{3}$ is equal to $\frac{4}{12}$

Fill in a fraction (with a value of 1) that will complete each operation.

1. $\frac{2}{5} = \frac{2 \times \boxed{3}}{5 \times \boxed{}} = \frac{6}{15}$ so   is equal to $\frac{\boxed{}}{\boxed{}}$

2. $\frac{3}{4} = \frac{3 \times \boxed{}}{4 \times \boxed{4}} = \frac{12}{16}$ so $\frac{\boxed{}}{\boxed{}}$ is equal to $\frac{\boxed{12}}{\boxed{16}}$

3. $\frac{5}{6} = \frac{5 \times \boxed{}}{6 \times \boxed{}} = \frac{10}{12}$ so $\frac{\boxed{}}{\boxed{}}$ is equal to $\frac{\boxed{}}{\boxed{}}$

4. $\frac{2}{7} = \frac{2 \times \boxed{}}{7 \times \boxed{}} = \frac{4}{14}$ so $\frac{\boxed{}}{\boxed{}}$ is equal to $\frac{\boxed{}}{\boxed{}}$

5. $\frac{1}{8} = \frac{1 \times \boxed{}}{8 \times \boxed{}} = \frac{3}{24}$ so $\frac{\boxed{}}{\boxed{}}$ is equal to $\frac{\boxed{}}{\boxed{}}$

6. $\frac{2}{9} = \frac{2 \times \boxed{}}{9 \times \boxed{}} = \frac{6}{27}$ so   is equal to $\frac{\boxed{}}{\boxed{}}$

Find the Numerators

Remember to multiply both the numerator and the denominator by the same number.

$$\frac{3}{4} = \frac{9}{12} \qquad \text{because} \qquad \frac{3}{4} = \frac{3 \times \boxed{3}}{4 \times \boxed{3}} = \frac{9}{12}$$

Fill in the correct numerators to make the fractions equal in value.

1. $\dfrac{1}{4} = \dfrac{}{20}$ because $\dfrac{1 \times \boxed{}}{4 \times } = \dfrac{}{20}$

2. $\dfrac{3}{5} = \dfrac{}{15}$ because $\dfrac{3 \times \boxed{}}{5 \times } = \dfrac{}{15}$

3. $\dfrac{4}{7} = \dfrac{}{21}$ because $\dfrac{4 \times \boxed{}}{7 \times } = \dfrac{}{21}$

4. $\dfrac{1}{3} = \dfrac{}{15}$ because $\dfrac{1 \times \boxed{}}{3 \times } = \dfrac{}{15}$

5. $\dfrac{5}{6} = \dfrac{}{12}$ because $\dfrac{5 \times \boxed{}}{6 \times } = \dfrac{}{12}$

6. $\dfrac{2}{9} = \dfrac{}{36}$ because $\dfrac{2 \times \boxed{}}{9 \times } = \dfrac{}{36}$

7. $\dfrac{2}{3} = \dfrac{}{18}$ because $\dfrac{2 \times \boxed{}}{3 \times } = \dfrac{}{18}$

8. $\dfrac{3}{4} = \dfrac{}{48}$ because $\dfrac{3 \times \boxed{}}{4 \times } = \dfrac{}{48}$

Practice

To find equivalent fractions, multiply the numerator and the denominator of the fraction by the same number.

Complete the fractions.

1. $\dfrac{1 \times 4}{3 \times 4} = \dfrac{4}{12}$

2. $\dfrac{3 \times 3}{4 \times 3} = \dfrac{9}{}$

3. $\dfrac{1}{2} = \dfrac{2}{}$

4. $\dfrac{2}{3} = \dfrac{}{9}$

5. $\dfrac{1}{4} = \dfrac{4}{}$

6. $\dfrac{1}{5} = \dfrac{}{45}$

7. $\dfrac{1}{3} = \dfrac{}{27}$

8. $\dfrac{1}{7} = \dfrac{2}{}$

9. $\dfrac{2}{5} = \dfrac{4}{}$

10. $\dfrac{1}{6} = \dfrac{}{42}$

11. $\dfrac{7}{10} = \dfrac{21}{}$

12. $\dfrac{5}{7} = \dfrac{35}{}$

13. $\dfrac{7}{12} = \dfrac{}{24}$

14. $\dfrac{11}{50} = \dfrac{}{100}$

15. $\dfrac{9}{20} = \dfrac{18}{}$

16. $\dfrac{7}{15} = \dfrac{14}{}$

17. $\dfrac{11}{14} = \dfrac{}{28}$

18. $\dfrac{7}{8} = \dfrac{}{32}$

19. $\dfrac{5}{9} = \dfrac{25}{}$

20. $\dfrac{1}{4} = \dfrac{}{12}$

Equivalent Fractions Review

Complete each problem.

1.

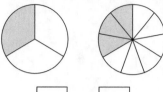

$$\frac{\boxed{}}{\boxed{}} = \frac{\boxed{}}{\boxed{}}$$

2. $\frac{1}{3} = \frac{}{6} = \frac{}{9} = \frac{4}{}$

3. $\frac{2}{5} = \frac{4}{} = \frac{}{15} = \frac{8}{}$

4. $\frac{3}{} = \frac{}{10} = \frac{9}{} = \frac{}{20}$

5. a) $\frac{5}{8} = \frac{}{24}$

 b) $\frac{5}{6} = \frac{}{24}$

 c) $\frac{5}{12} = \frac{}{24}$

6. a) $\frac{1}{2} = \frac{}{18}$

 b) $\frac{3}{6} = \frac{}{18}$

 c) $\frac{4}{9} = \frac{}{18}$

7. a) $\frac{2}{3} = \frac{}{15}$

 b) $\frac{1}{5} = \frac{}{15}$

 c) $\frac{3}{15} = \frac{}{30}$

8. a) $\frac{5}{7} = \frac{15}{}$

 b) $\frac{3}{4} = \frac{12}{}$

 c) $\frac{1}{8} = \frac{5}{}$

Less Than or Greater Than

Find the missing numerators and then compare the fractions using the symbols < (less than) or > (greater than).

1. a) $\dfrac{1}{4}$ = $\dfrac{2}{8}$

b) $\dfrac{1}{2}$ = $\dfrac{4}{8}$

c) $\dfrac{1}{4}$ $\boxed{<}$ $\dfrac{1}{2}$

$\left(\dfrac{2}{8}\right)$ $\left(\dfrac{4}{8}\right)$

4. a) $\dfrac{2}{3}$ = $\dfrac{}{15}$

b) $\dfrac{4}{5}$ = $\dfrac{}{15}$

c) $\dfrac{2}{3}$ $\bigcirc$ $\dfrac{4}{5}$

2. a) $\dfrac{1}{3}$ = $\dfrac{}{24}$

b) $\dfrac{3}{8}$ = $\dfrac{}{24}$

c) $\dfrac{1}{3}$ $\bigcirc$ $\dfrac{3}{8}$

5. a) $\dfrac{1}{2}$ = $\dfrac{}{6}$

b) $\dfrac{2}{3}$ = $\dfrac{}{6}$

c) $\dfrac{1}{2}$ $\bigcirc$ $\dfrac{2}{3}$

3. a) $\dfrac{5}{6}$ = $\dfrac{}{24}$

b) $\dfrac{3}{4}$ = $\dfrac{}{24}$

c) $\dfrac{5}{6}$ $\bigcirc$ $\dfrac{3}{4}$

6. a) $\dfrac{3}{4}$ = $\dfrac{}{12}$

b) $\dfrac{4}{6}$ = $\dfrac{}{12}$

c) $\dfrac{3}{4}$ $\bigcirc$ $\dfrac{4}{6}$

Compare Using the Number Line

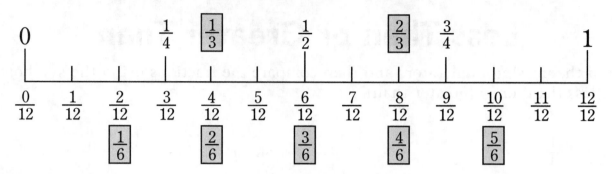

Find both fractions on the number line and compare using the symbols
< (less than), > (greater than), or = (equal to). Remember: the closer a fraction
is to zero, the smaller its value.

1. $\frac{1}{4}$ ⟨<⟩ $\frac{1}{3}$

2. $\frac{4}{6}$ ◯ $\frac{2}{3}$

3. $\frac{5}{6}$ ◯ $\frac{7}{12}$

4. $\frac{11}{12}$ ◯ $\frac{3}{4}$

5. $\frac{5}{12}$ ◯ $\frac{7}{12}$

6. $\frac{6}{12}$ ◯ $\frac{3}{6}$

7. $\frac{5}{6}$ ◯ $\frac{5}{12}$

8. $\frac{1}{2}$ ◯ $\frac{7}{12}$

9. $\frac{1}{6}$ ◯ $\frac{2}{12}$

10. $\frac{3}{4}$ ◯ $\frac{8}{12}$

11. $\frac{2}{6}$ ◯ $\frac{1}{12}$

12. $\frac{9}{12}$ ◯ $\frac{5}{6}$

Connect Tags to Lines

Connect each tag to the number line. Begin by counting the number of parts each line is divided into. You may need to use the number line on page 36 to find the equivalent fractions. Write the correct letter below each tag.

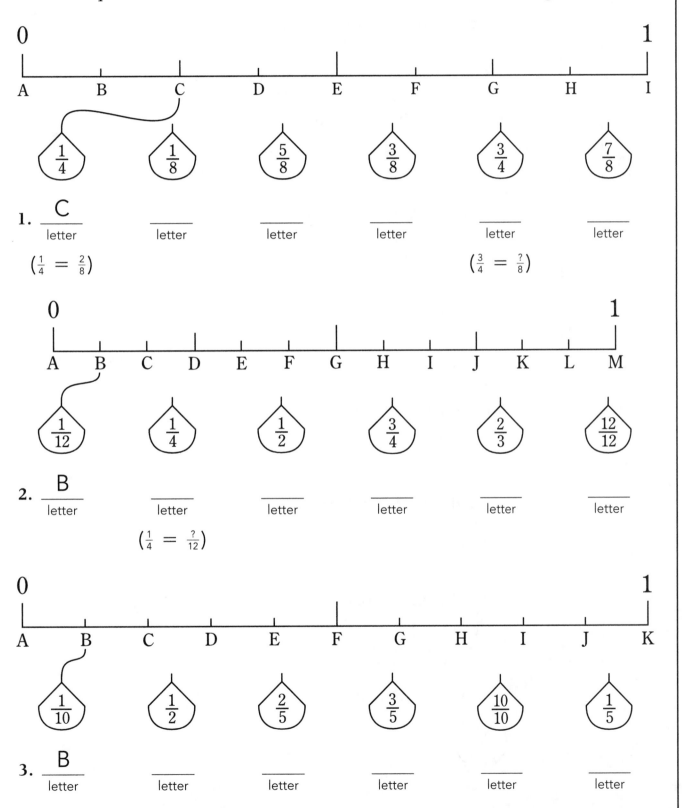

0   1

A   B   C   D   E   F   G   H   I

$\frac{1}{4}$ $\frac{1}{8}$ $\frac{5}{8}$ $\frac{3}{8}$ $\frac{3}{4}$ $\frac{7}{8}$

1. C
    letter   letter   letter   letter   letter   letter

$(\frac{1}{4} = \frac{2}{8})$ $(\frac{3}{4} = \frac{?}{8})$

0   1

A   B   C   D   E   F   G   H   I   J   K   L   M

$\frac{1}{12}$ $\frac{1}{4}$ $\frac{1}{2}$ $\frac{3}{4}$ $\frac{2}{3}$ $\frac{12}{12}$

2. B
    letter   letter   letter   letter   letter   letter

$(\frac{1}{4} = \frac{?}{12})$

0   1

A   B   C   D   E   F   G   H   I   J   K

$\frac{1}{10}$ $\frac{1}{2}$ $\frac{2}{5}$ $\frac{3}{5}$ $\frac{10}{10}$ $\frac{1}{5}$

3. B
    letter   letter   letter   letter   letter   letter

Fractions Show Comparisons

A fraction can be thought of as a comparison. Write the fraction for each drawing below.

Fraction

1. $\dfrac{3}{5}$

Shaded squares to total squares

2. ___

Indicated units to total units

3. ___

Shaded circle to total circles

4. ___

Shaded part to total

5. ___

Shaded part to total

6. ___

Shaded parts to total

7. ___

Shaded boxes to total boxes

8. ___

Shaded part to total

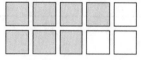

More Comparisons

A fraction can be thought of as a comparison. For example, a basketball team can compare its number of wins to the total number of games played. If a team plays 10 games and wins 7, the fraction of wins to total games would be:

$$\frac{\text{part}}{\text{total}} = \frac{7}{10} \text{ of the games were won}$$

Write these comparisons as fractions.

1. 30 students
   15 earned an "A" in math

   $$\frac{\text{part}}{\text{total}} = \frac{\boxed{}}{30} \text{ received an "A"}$$

2. 10 hits
   20 times at bat

   $$\frac{\text{part}}{\text{total}} = \frac{\boxed{}}{\boxed{}} \text{ were hits}$$

3. 10 high dives
   7 perfect scores

   $$\frac{\text{part}}{\text{total}} = \frac{\boxed{}}{\boxed{}} \text{ perfect scores}$$

4. Joan's book has 125 pages.
   She read 25 pages.

   $$\frac{\text{part}}{\text{total}} = \frac{\boxed{}}{\boxed{}} \text{ of the book was read}$$

5. 6 coins
   2 are pennies

   $$\frac{\text{part}}{\text{total}} = \frac{\boxed{}}{\boxed{}} \text{ of the coins are pennies}$$

6. A store sold 10 radios.
   2 were returned.

   $$\frac{\text{part}}{\text{total}} = \frac{\boxed{}}{\boxed{}} \text{ of the radios were returned}$$

7. The cake was divided into 12 equal pieces. We ate 2 of them.

   $$\frac{\text{part}}{\text{total}} = \frac{\boxed{}}{\boxed{}} \text{ of the cake was eaten}$$

8. It rained 3 out of 4 days.

   $$\frac{\text{part}}{\text{total}} = \frac{\boxed{}}{\boxed{}} \text{ of the days were rainy}$$

Comparisons Review

1. Which is a correct drawing for $\frac{1}{4}$?

  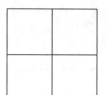

    A   B   C

Answer: _____

6. Shade $\frac{5}{4}$ which equals _____
                                 mixed number

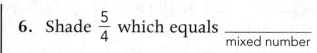

2. Shade $\frac{4}{5}$

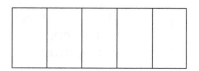

7. $\frac{3}{4} = \frac{}{20}$

3. Which is larger $\frac{3}{5}$ or $\frac{3}{8}$?

Answer: _____

8. $\frac{5}{8} = \frac{10}{}$

4. Shade $\frac{3}{4}$ of 8

9. 4 times at bat
1 hit

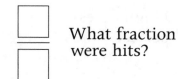
What fraction were hits?

5. Circle $\frac{1}{3}$ of 9

10. What fraction of the line is shown at point A?

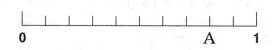

Answer: _____

Divisibility Rule for 2

Sometimes it is helpful to know if a number is divisible by 2.

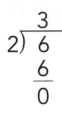

6 **is** divisible by 2 because it divides exactly with a remainder of 0.

5 **is not** divisible by 2 because it does not divide exactly. The remainder is not 0.

Any number that ends in 0, 2, 4, 6 or 8 is divisible by 2.

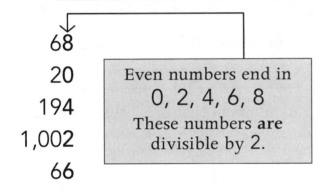

68
20
194
1,002
66

Even numbers end in
0, 2, 4, 6, 8
These numbers **are** divisible by 2.

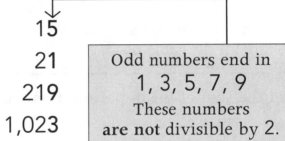

15
21
219
1,023
17

Odd numbers end in
1, 3, 5, 7, 9
These numbers **are not** divisible by 2.

Circle the numbers that are divisible by 2.

1. 22
2. 71
3. 15
4. 293
5. 96
6. 104

7. 107
8. 39
9. 75
10. 212
11. 200
12. 93

13. 655
14. 938
15. 1,004
16. 86
17. 95
18. 100

Divisibility Rule for 3

Sometimes it is helpful to know if a number is divisible by 3.

$$3\overline{)12} = 4$$
$$\underline{12}$$
$$0$$

12 **is** divisible by 3 because it divides exactly with a remainder of 0.

$$3\overline{)11} = 3$$
$$\underline{9}$$
$$2$$

11 **is not** divisible by 3 because it does not divide exactly. The remainder is not 0.

If the sum of the digits in any number is divisible by 3, the number is divisible by 3.

$$3\overline{)\boxed{5\ 4}}$$
5 + 4 = 9

$$3\overline{)\boxed{6\ 4\ 2}}$$
6 + 4 + 2 = 12

$$3\overline{)\boxed{5\ 6\ 7}}$$
5 + 6 + 7 = 18

The sums of the digits (9, 12, and 18) are all divisible by 3, so 54, 642, and 567 are all divisible by 3.

Find the sum of the digits and then circle the numbers that are divisible by 3.

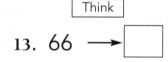

1. (27) → 9	7. 46 → ☐	13. 66 → ☐
2. 35 → ☐	8. 57 → ☐	14. 54 → ☐
3. 60 → ☐	9. 21 → ☐	15. 621 → ☐
4. 56 → ☐	10. 24 → ☐	16. 304 → ☐
5. 92 → ☐	11. 19 → ☐	17. 555 → ☐
6. 31 → ☐	12. 85 → ☐	18. 206 → ☐

Think: 2 + 7 =

Divisibility Rule for 5

Sometimes it is helpful to know if a number is divisible by 5.

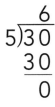

        6
    5)3 0
      3 0
      ───
        0

30 **is** divisible by 5 because it divides exactly with a remainder of 0.

        5
    5)2 9
      2 5
      ───
        4

29 **is not** divisible by 5 because it does not divide exactly. The remainder is not 0.

Any number that ends in 0 or 5 is divisible by 5.

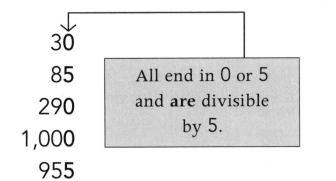

30
85
290
1,000
955

All end in 0 or 5 and **are** divisible by 5.

36
82
41
109
164

These numbers **do not** end in 0 or 5 so are **not** divisible by 5.

Circle all the numbers that are divisible by 5.

1. 48

2. (55)

3. 62

4. 93

5. 15

6. 50

7. 62

8. 35

9. 80

10. 572

11. 24

12. 175

13. 180

14. 105

15. 69

16. 908

17. 370

18. 90

Divisibility Rule for 10

Sometimes it is helpful to know if a number is divisible by 10.

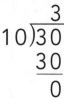

30 **is** divisible by 10 because it divides exactly with a remainder of 0.

38 **is not** divisible by 10 because it does not divide exactly. The remainder is not 0.

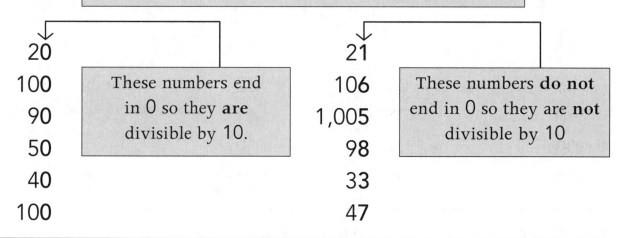

Any number that ends in 0 is divisible by 10.

20	These numbers end in 0 so they **are** divisible by 10.
100	
90	
50	
40	
100	

21	These numbers **do not** end in 0 so they are **not** divisible by 10
106	
1,005	
98	
33	
47	

Circle all the numbers that are divisible by 10.

1. 30
2. 38
3. 92
4. 57
5. 65
6. 105

7. 638
8. 70
9. 815
10. 309
11. 88
12. 905

13. 556
14. 200
15. 681
16. 93
17. 755
18. 450

Divisibility Practice

Check the column if the number is divisible by 2, 3, 5, or 10.

	Number	Divisible by 2	Divisible by 3	Divisible by 5	Divisible by 10
1.	40	✓		✓	✓
2.	144				
3.	94				
4.	540				
5.	1,000				
6.	29				
7.	45				
8.	85				
9.	342				
10.	70				
11.	100				
12.	65				
13.	38				
14.	153				
15.	384				
16.	89				
17.	420				
18.	5,546				

Finding Factors

Use factors to simplify fractions.

Factors are:
- all the numbers that can be multiplied to find a given number
- the number itself and 1

To find all the factors of 15, **think:**

$$\underset{\text{factor}}{1} \times \underset{\text{factor}}{15} = \underset{\text{product}}{15}$$

$$\underset{\text{factor}}{3} \times \underset{\text{factor}}{5} = \underset{\text{product}}{15}$$

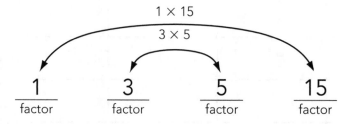

1. List all the factors for 15. __1__,_____,_____,__15__

2. How many factors does 15 have? _____

To find all the factors of 20, **think:**

$$1 \times 20 = 20$$
$$2 \times 10 = 20$$
$$4 \times 5 = 20$$

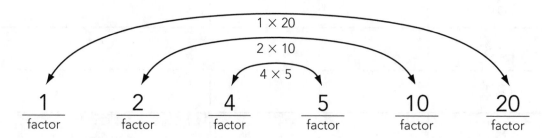

3. List all the factors for 20. __1__,_____,_____,_____,_____,__20__

4. How many factors does 20 have? _____

Name the Factors

Find all the factors of 18.

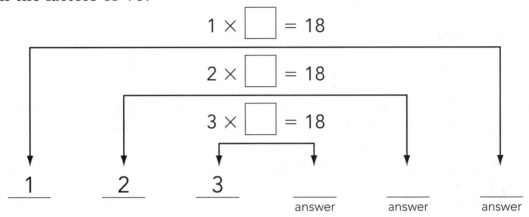

A. ___1___   ___2___   ___3___   _____   _____   _____
                                          answer   answer   answer

B.   How many factors does 18 have? _____
                                        answer

Write all of the factors for each number.

1. Name all the factors for 12: _____,__2__,_____,_____,_____,__12__

                                        list only once

2. Name all the factors for 9: _____,__3__,_____

3. Name all the factors for 10: _____,_____,_____,_____

4. Name all the factors for 28: _____,_____,_____,_____,_____,_____

5. Name all the factors for 8: _____,_____,_____,_____

6. Name all the factors for 4: _____,_____,_____

7. Name all the factors for 7: _____,_____

8. Name all the factors for 16: _____,_____,_____,_____,_____

9. Name all the factors for 24: _____,_____,_____,_____,_____,_____,_____,_____

10. Name all the factors for 30: _____,_____,_____,_____,_____,_____,_____,_____

Greatest Common Factor

Find the greatest common factor (GCF) for 8 and 12.

The factors of 8 are: 1 2 4 8

The factors of 12 are: 1 2 3 4 6 12

The factors 1, 2, and 4 are factors of both 8 and 12 and are called common factors. The number 4 is the greatest common factor of 8 and 12.

> The largest factor that two or more numbers have in common is the **greatest common factor.**

Find the greatest common factor (GCF).

1. 4: _____,_____,_____
   factors of 4

2. 8: _____,_____,_____,_____
   factors of 8

3. Common factors of 4 and 8: _____,_____,_____

4. Greatest common factor of 4 and 8: _____

5. 6: _____,_____,_____,_____
   factors of 6

6. 9: _____,_____,_____
   factors of 9

7. Common factors of 6 and 9: _____,_____

8. Greatest common factor of 6 and 9: _____

9. 16: _____,_____,_____,_____,_____
   factors of 16

10. 20: _____,_____,_____,_____,_____,_____
    factors of 20

11. Common factors of 16 and 20: _____,_____,_____

12. Greatest common factor of 16 and 20: _____

Find the Greatest Common Factor

Find the greatest common factor (GCF).

1. 10: _____,_____,_____,_____
   factors of 10

2. 15: _____,_____,_____,_____
   factors of 15

3. Common factors of 10 and 15: _____,_____

4. Greatest common factor of 10 and 15: _____

5. 8: _____,_____,_____,_____
   factors of 8

6. 12: _____,_____,_____,_____,_____,_____
   factors of 12

7. Common factors of 8 and 12: _____,_____,_____

8. Greatest common factor of 8 and 12: _____

9. 14: _____,_____,_____,_____
   factors of 14

10. 24: _____,_____,_____,_____,_____,_____,_____,_____
    factors of 24

11. Common factors of 14 and 24: _____,_____

12. Greatest common factor of 14 and 24: _____

13. 21: _____,_____,_____,_____
    factors of 21

14. 36: _____,_____,_____,_____,_____,_____,_____,_____,_____
    factors of 36

15. Common factors of 21 and 36: _____,_____

16. Greatest common factor of 21 and 36: _____

Factors with Fractions

You many need to find common factors between the numerator and the denominator of a fraction.

You don't always need to make a list of factors. Think of all whole numbers that divide exactly into both the numerator and the denominator. These numbers will be the common factors.

$\frac{6}{12}$

A. The common factors of 6 and 12 are: ___1___,_____,_____,_____

B. The greatest common factor (GCF) is: _____

Find the greatest common factor for the numerator and denominator in each fraction.

$\frac{6}{15}$

1. The common factors of 6 and 15 are: _____,_____

2. The greatest common factor (GCF) is: _____

$\frac{10}{30}$

3. The common factors of 10 and 30 are: _____,_____,_____,_____

4. The greatest common factor (GCF) is: _____

$\frac{10}{16}$

5. The common factors of 10 and 16 are: _____,_____

6. The greatest common factor (GCF) is: _____

$\frac{8}{16}$

7. The common factors of 8 and 16 are: _____,_____,_____,_____

8. The greatest common factor (GCF) is: _____

$\frac{7}{21}$

9. The common factors of 7 and 21 are: _____,_____

10. The greatest common factor (GCF) is: _____

Apply Your Skills

Find the greatest common factor (GCF) for each fraction.

1. $\frac{6}{12}$ GCF = ____

2. $\frac{3}{15}$ GCF = ____

3. $\frac{4}{6}$ GCF = ____

4. $\frac{5}{30}$ GCF = ____

5. $\frac{2}{14}$ GCF = ____

6. $\frac{14}{21}$ GCF = ____

7. $\frac{8}{24}$ GCF = ____

8. $\frac{9}{36}$ GCF = ____

9. $\frac{8}{10}$ GCF = ____

10. $\frac{6}{15}$ GCF = ____

11. $\frac{2}{4}$ GCF = ____

12. $\frac{6}{24}$ GCF = ____

13. $\frac{3}{36}$ GCF = ____

14. $\frac{12}{15}$ GCF = ____

15. $\frac{9}{12}$ GCF = ____

16. $\frac{15}{18}$ GCF = ____

17. $\frac{20}{30}$ GCF = ____

18. $\frac{25}{40}$ GCF = ____

19. $\frac{5}{30}$ GCF = ____

20. $\frac{9}{15}$ GCF = ____

21. $\frac{9}{18}$ GCF = ____

22. $\frac{12}{36}$ GCF = ____

23. $\frac{11}{22}$ GCF = ____

24. $\frac{14}{35}$ GCF = ____

Simplify Fractions

Sometimes you need to simplify fractions or reduce them to their lowest terms.

> To simplify a fraction, divide both the numerator and the denominator by the greatest common factor.

$\dfrac{6}{9} \div \dfrac{3}{3} = \dfrac{2}{3}$ Think: GCF of 6 and 9 is 3.

$\dfrac{4}{20} \div \dfrac{4}{4} = \dfrac{1}{5}$ Think: GCF of 4 and 20 is 4.

Find the greatest common factor and simplify.

1. $\dfrac{14}{49}$ GCF = ___7___

   $\dfrac{14}{49} \div \dfrac{7}{7} = $ _____ finish

2. $\dfrac{4}{12}$ GCF = _____

   $\dfrac{4}{12} \div \dfrac{}{} = $ _____ simplified fraction

3. $\dfrac{24}{32}$ GCF = _____

   _____ simplified fraction

4. $\dfrac{20}{30}$ GCF = _____

   _____ simplified fraction

5. $\dfrac{11}{33}$ GCF = _____

   _____ simplified fraction

6. $\dfrac{18}{45}$ GCF = _____

   _____ simplified fraction

7. $\dfrac{25}{30}$ GCF = _____

   _____ simplified fraction

8. $\dfrac{30}{45}$ GCF = _____

   _____ simplified fraction

9. $\dfrac{32}{36}$ GCF = _____

   _____ simplified fraction

10. $\dfrac{26}{39}$ GCF = _____

    _____ simplified fraction

Simplify

Find the greatest common factor and simplify each fraction.

1. $\dfrac{6}{12} \div \dfrac{6}{6} = \dfrac{1}{2}$

   GCF = 6

2. $\dfrac{3}{15} \div \dfrac{\boxed{}}{} = \dfrac{1}{5}$

3. $\dfrac{4}{6} \div \dfrac{2}{2} = \underline{}$

4. $\dfrac{5}{30} \div \dfrac{\boxed{}}{} = \underline{}$

5. $\dfrac{2}{14} \div \dfrac{\boxed{}}{} = \underline{}$

6. $\dfrac{14}{21} \div \dfrac{\boxed{}}{} = \underline{}$

7. $\dfrac{8}{24} \div \dfrac{\boxed{}}{} = \underline{}$

8. $\dfrac{9}{36} \div \dfrac{\boxed{}}{} = \underline{}$

9. $\dfrac{8}{10} \div \dfrac{2}{2} = \underline{}$

10. $\dfrac{6}{15} \div \dfrac{\boxed{}}{} = \underline{}$

11. $\dfrac{2}{4} \div \dfrac{\boxed{}}{} = \underline{}$

12. $\dfrac{6}{24} \div \dfrac{\boxed{}}{} = \underline{}$

13. $\dfrac{3}{36} \div \dfrac{\boxed{}}{} = \underline{}$

14. $\dfrac{12}{15} \div \dfrac{\boxed{}}{} = \underline{}$

15. $\dfrac{9}{12} \div \dfrac{\boxed{}}{} = \underline{}$

16. $\dfrac{15}{18} \div \dfrac{\boxed{}}{} = \underline{}$

17. $\dfrac{20}{30} \div \dfrac{10}{10} = \underline{}$

18. $\dfrac{25}{40} \div \dfrac{\boxed{}}{} = \underline{}$

19. $\dfrac{5}{50} \div \dfrac{\boxed{}}{} = \underline{}$

20. $\dfrac{9}{15} \div \dfrac{\boxed{}}{} = \underline{}$

21. $\dfrac{9}{18} \div \dfrac{\boxed{}}{} = \underline{}$

22. $\dfrac{12}{36} \div \dfrac{\boxed{}}{} = \underline{}$

23. $\dfrac{11}{22} \div \dfrac{\boxed{}}{} = \underline{}$

24. $\dfrac{14}{35} \div \dfrac{\boxed{}}{} = \underline{}$

Simplest Form

A fraction is in its simplest form when the greatest common factor (GCF) is 1.

$$\frac{6}{8} \div \frac{2}{2} = \frac{3}{4}$$

GCF of 6 and 8 is 2.
6: 1, 2, 3, 6
8: 1, 2, 4, 8

$$\frac{5}{8} \div \frac{1}{1} = \frac{5}{8}$$ already simplified

GCF of 5 and 8 is 1.
5: 1, 5
8: 1, 2, 4, 8

Simplify when necessary. If the fraction **does not** need simplifying, write "simplified."

1. $\frac{3}{5}$ = simplified

2. $\frac{15}{18}$ = $\frac{5}{6}$

3. $\frac{3}{16}$ =

4. $\frac{9}{18}$ =

5. $\frac{4}{6}$ =

6. $\frac{7}{8}$ =

7. $\frac{15}{16}$ =

8. $\frac{9}{12}$ =

9. $\frac{7}{10}$ =

10. $\frac{1}{8}$ =

11. $\frac{3}{9}$ =

12. $\frac{4}{5}$ =

13. $\frac{16}{20}$ =

14. $\frac{17}{18}$ =

15. $\frac{3}{8}$ =

16. $\frac{4}{9}$ =

17. $\frac{9}{15}$ =

18. $\frac{3}{4}$ =

19. $\frac{6}{8}$ =

20. $\frac{8}{8}$ =

21. $\frac{8}{12}$ =

22. $\frac{12}{16}$ =

23. $\frac{5}{8}$ =

24. $\frac{10}{12}$ =

Think It Through

Sometimes you might not be able to think of the greatest common factor, but you can come up with a common factor.

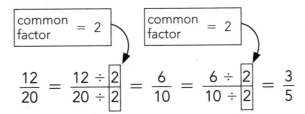

If you cannot think of the greatest common factor, keep dividing by common factors until the fraction is simplified to its lowest terms. But remember: finding the greatest common factor will help you simplify a problem quickly.

Simplify the fractions. Use more than one common factor if necessary.

1. $\dfrac{48}{60} =$

2. $\dfrac{26}{78} =$

3. $\dfrac{45}{60} =$

4. $\dfrac{18}{30} =$

5. $\dfrac{21}{63} =$

6. $\dfrac{12}{30} =$

7. $\dfrac{15}{30} =$

8. $\dfrac{16}{32} =$

9. $\dfrac{22}{44} =$

10. $\dfrac{24}{36} =$

11. $\dfrac{12}{18} =$

12. $\dfrac{16}{40} =$

Lowest Terms

Change each fraction to its simplest form. If the fraction does not need simplifying, write "simplified."

1. $\dfrac{16}{48}$ =

2. $\dfrac{6}{24}$ =

3. $\dfrac{9}{27}$ =

4. $\dfrac{9}{36}$ =

5. $\dfrac{18}{54}$ =

6. $\dfrac{24}{36}$ =

7. $\dfrac{9}{12}$ =

8. $\dfrac{7}{14}$ =

9. $\dfrac{16}{32}$ =

10. $\dfrac{4}{12}$ =

11. $\dfrac{3}{9}$ =

12. $\dfrac{8}{24}$ =

13. $\dfrac{15}{28}$ =

14. $\dfrac{10}{12}$ =

15. $\dfrac{21}{42}$ =

16. $\dfrac{12}{16}$ =

17. $\dfrac{14}{20}$ =

18. $\dfrac{8}{16}$ =

19. $\dfrac{6}{21}$ =

20. $\dfrac{8}{18}$ =

21. $\dfrac{15}{21}$ =

22. $\dfrac{13}{24}$ =

23. $\dfrac{15}{18}$ =

24. $\dfrac{12}{14}$ =

25. $\dfrac{15}{24}$ =

26. $\dfrac{8}{12}$ =

27. $\dfrac{10}{35}$ =

28. $\dfrac{7}{20}$ =

29. $\dfrac{15}{45}$ =

30. $\dfrac{30}{42}$ =

Apply Your Skills

Simplify the fractions.

1. $\dfrac{54}{60} =$

Answer: _____

2. $\dfrac{11}{44} =$

Answer: _____

3. $\dfrac{28}{56} =$

Answer: _____

4. $\dfrac{7}{21} =$

Answer: _____

5. $\dfrac{9}{72} =$

Answer: _____

6. $\dfrac{8}{20} =$

Answer: _____

7. $\dfrac{3}{39} =$

Answer: _____

8. $\dfrac{27}{81} =$

Answer: _____

Divisibility and Common Factors Review

1. Circle all numbers that are divisible by 3.

63	16	183	9
126	23	17	249
32	34	514	801

2. Circle all numbers that are divisible by 5.

155	220	312	118
100	550	345	1,000
214	335	126	980

3. Name all the factors for 4:

_____ , _____ , _____

4. Name all the factors for 10:

_____ , _____ , _____ , _____

5. Find the GCF for 8 and 24:

GCF = _____

6. $\frac{6}{18}$

GCF = _____

7. $\frac{14}{35}$

GCF = _____

Simplify the fractions.

8. a) $\frac{16}{20} =$ b) $\frac{3}{9} =$

9. a) $\frac{15}{45} =$ b) $\frac{3}{12} =$

10. a) $\frac{12}{18} =$ b) $\frac{24}{78} =$

Changing Fractions to Decimals

Often it is necessary to change a fraction to a decimal. Think of the fraction bar as a division bar.

Change $\frac{1}{4}$ to a decimal.

Step 1: $\frac{1}{4}$ ◄── division bar   Think: $1 \div 4$

Step 2: $\frac{1}{4}$ ⟍ = $\begin{array}{r} .25 \\ 4\overline{)1.00} \\ \underline{8} \\ 20 \\ \underline{20} \\ 0 \end{array}$ so $\frac{1}{4} = .25$

Add as many zeros as you need until the division comes out exactly. The remainder will be zero.

Rewrite each fraction as a division problem.

1. $\frac{1}{5}$ ⟍ _____ $5\overline{)1.0}$

2. $\frac{3}{4}$

3. $\frac{3}{8}$

4. $\frac{1}{8}$ ⟍ _____ $8\overline{)1.000}$

5. $\frac{3}{5}$

6. $\frac{7}{8}$

More Practice: Fractions to Decimals

Find the decimal value of these fractions. Remember to write as many zeros as necessary to have a remainder of zero.

1. $\frac{10}{8}$ = __.___

$$
\begin{array}{r}
1.25 \\
8\overline{)10.00} \\
\underline{8} \\
20 \\
\underline{16} \\
40 \\
\underline{40} \\
0
\end{array}
$$

2. $\frac{18}{15}$ =

3. $\frac{9}{12}$ =

4. $\frac{1}{2}$ =

5. $\frac{5}{8}$ = .___

$$8\overline{)5.000}$$

6. $\frac{8}{10}$ =

7. $\frac{22}{16}$ =

8. $\frac{15}{4}$ =

9. $\frac{3}{12}$ = .__

$$12\overline{)3}$$

10. $\frac{24}{10}$ =

11. $\frac{14}{16}$ =

12. $\frac{11}{5}$ =

Working with Remainders

Some decimal forms of fractions have remainders after you have divided to the hundredths place.

- Divide to the hundredths place.
- Write all remainders as fractions: put the remainder over the original denominator.

$$\frac{1}{3} = 3\overline{)1.00}\;.33\tfrac{1}{3}$$ ← put the remainder over the denominator

$$\begin{array}{r} .33\tfrac{1}{3} \\ 3\overline{)1.00} \\ \underline{9} \\ 10 \\ \underline{9} \\ 1 \end{array}$$ ← remainder

$$\frac{5}{8} = 8\overline{)5.00}\;.62\tfrac{4}{8} = .62\tfrac{1}{2}$$ ← simplify the remainder

$$\begin{array}{r} .62\tfrac{4}{8} \\ 8\overline{)5.00} \\ \underline{48} \\ 20 \\ \underline{16} \\ 4 \end{array}$$ ← remainder

---

Find the decimal form of each fraction. Divide to the hundredths place and write all remainders as fractions.

1. $\frac{2}{3}$

$$\begin{array}{r} .6 \\ 3\overline{)2.00} \\ \underline{18} \\ 20 \\ \overline{} \\ \overline{} \end{array}$$

2. $\frac{5}{6}$

$$6\overline{)5.00}$$

3. $\frac{7}{8}$

4. $\frac{1}{9}$

5. $\frac{2}{9}$

6. $\frac{4}{9}$

Practice Helps

- Carry out all decimals to the hundredths place (two decimal places).
- Change all remainders to fractions.

$$\frac{10}{7} = 7\overline{)10.00} \quad 1.42\frac{6}{7}$$

$$\begin{array}{r} \underline{7} \\ 30 \\ \underline{28} \\ 20 \\ \underline{14} \\ 6 \end{array}$$

← put the remainder over the denominator

← remainder

Find the decimal form of each fraction. Divide to the hundredths place and write all remainders as fractions.

1. $\frac{8}{18}$

$$18\overline{)8.00} \quad .4$$
$$\underline{72}$$

finish

2. $\frac{15}{7}$

3. $\frac{6}{7}$

4. $\frac{3}{27}$

5. $\frac{33}{21}$

6. $\frac{15}{9}$

Fractions and Decimals Review

Find the decimal value of these fractions.

Find the decimal form of each fraction. Divide to the hundredths place and write all remainders as fractions.

1. $\dfrac{4}{8}$ =

Answer: _____

5. $\dfrac{4}{10}$ =

Answer: _____

2. $\dfrac{3}{4}$ =

Answer: _____

6. $\dfrac{5}{8}$ =

Answer: _____

3. $\dfrac{11}{5}$ =

Answer: _____

7. $\dfrac{15}{6}$ =

Answer: _____

4. $\dfrac{15}{4}$ =

Answer: _____

8. $\dfrac{18}{5}$ =

Answer: _____

Half-Inch Measurements

Find the measurement in inches for each line.

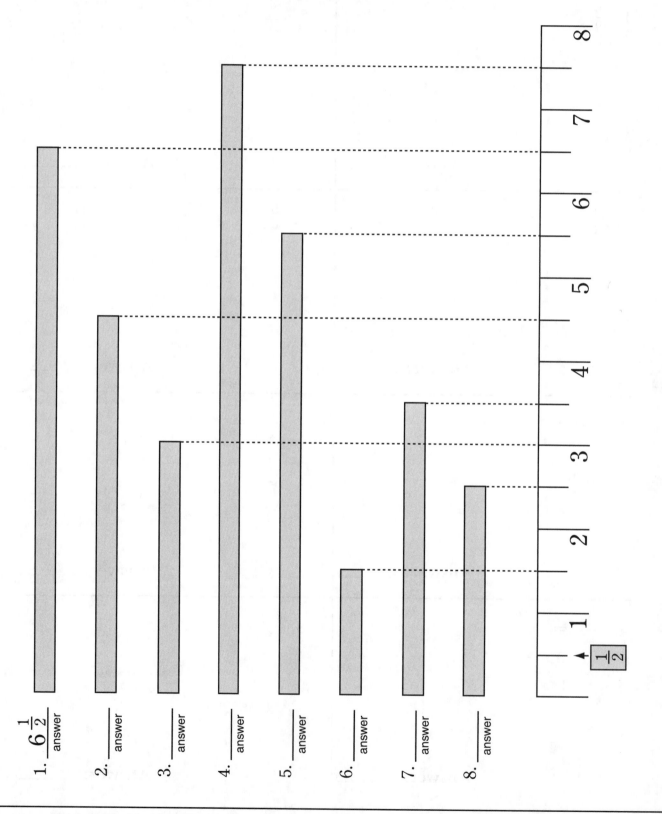

1. $6\frac{1}{2}$ ___ answer

2. ___ answer

3. ___ answer

4. ___ answer

5. ___ answer

6. ___ answer

7. ___ answer

8. ___ answer

Quarter-Inch Measurements

Find the measurements in inches for each line.

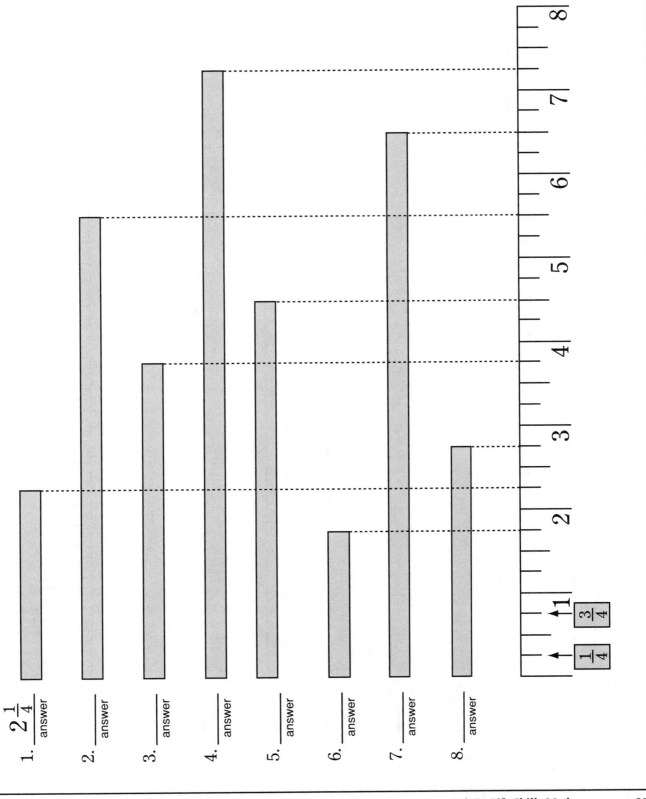

1. $2\frac{1}{4}$ answer
2. answer
3. answer
4. answer
5. answer
6. answer
7. answer
8. answer

Eighth-Inch Measurements

Find the measurement in inches for each line.

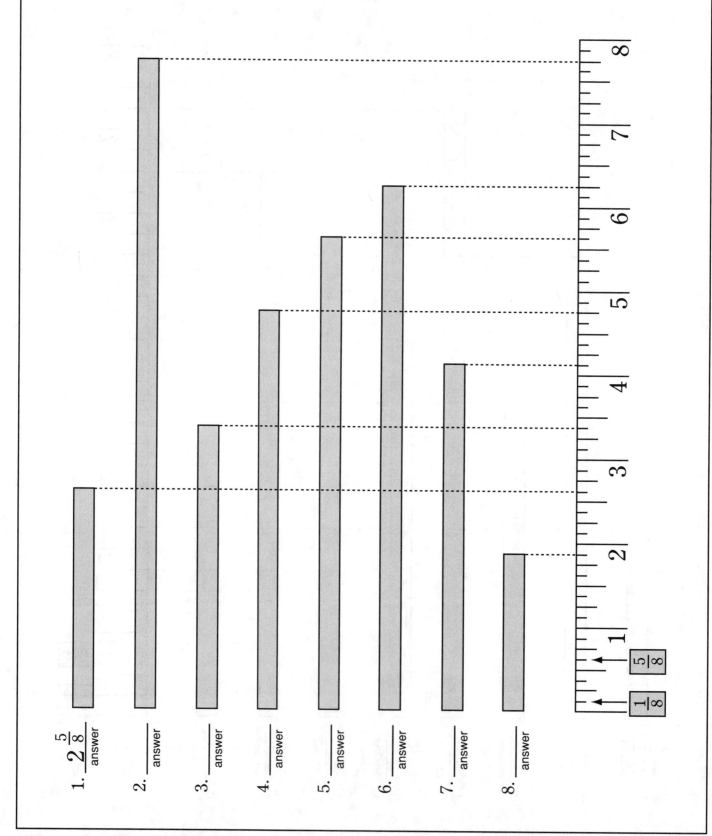

1. 2 5/8 answer
2. _____ answer
3. _____ answer
4. _____ answer
5. _____ answer
6. _____ answer
7. _____ answer
8. _____ answer

Draw the Measurements

Draw lines for the following lengths.

1. $5\frac{1}{4}$ inches

2. $1\frac{1}{2}$ inches

3. $4\frac{3}{4}$ inches

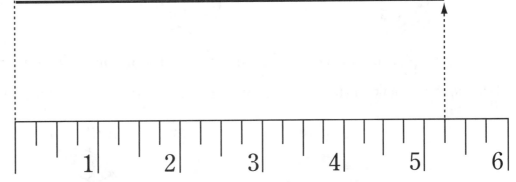

4. $3\frac{1}{4}$ inches

5. $5\frac{3}{4}$ inches

6. $1\frac{5}{8}$ inches

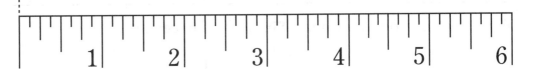

7. $5\frac{7}{8}$ inches

8. $2\frac{3}{8}$ inches

9. $4\frac{9}{16}$ inches

10. $4\frac{5}{16}$ inches

11. $1\frac{1}{4}$ inches

12. $3\frac{1}{2}$ inches

Money as Fractions

Penny

$1 \text{ cent} = \dfrac{1}{100}$ of a dollar

100 pennies make a dollar

Nickel

$5 \text{ cents} = \dfrac{1}{20}$ of a dollar

20 nickels make a dollar

Dime

$10 \text{ cents} = \dfrac{1}{10}$ of a dollar

10 dimes make a dollar

Quarter

$25 \text{ cents} = \dfrac{1}{4}$ of a dollar

4 quarters make a dollar

Half-Dollar

$50 \text{ cents} = \dfrac{1}{2}$ of a dollar

2 half-dollars make a dollar

Find the fraction of a dollar. Simplify if necessary.

1. 4 dimes = $\dfrac{4}{10}$ or $\dfrac{2}{5}$ of a dollar

6. 25 pennies = $\dfrac{\Box}{\Box}$ or $\dfrac{\Box}{\Box}$ of a dollar

2. 10 pennies = $\dfrac{10}{100}$ or $\dfrac{\Box}{\Box}$ of a dollar

7. 10 nickels = $\dfrac{\Box}{\Box}$ or $\dfrac{\Box}{\Box}$ of a dollar

3. 5 nickels = $\dfrac{\Box}{20}$ or $\dfrac{\Box}{\Box}$ of a dollar

8. 7 nickels = $\dfrac{\Box}{\Box}$ of a dollar

4. 3 quarters = $\dfrac{\Box}{\Box}$ of a dollar

9. 1 half-dollar = $\dfrac{\Box}{\Box}$ of a dollar

5. 5 dimes = $\dfrac{\Box}{\Box}$ or $\dfrac{\Box}{\Box}$ of a dollar

10. 8 dimes = $\dfrac{\Box}{\Box}$ or $\dfrac{\Box}{\Box}$ of a dollar

What Coin Is Missing?

Find the missing coin that completes each fraction of a dollar.

1.

$\frac{1}{4}$ of 1 dollar

<u>25</u> cents

2.

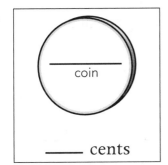

25 cents   <u>5</u> cents

$\frac{3}{10}$ of 1 dollar

(Hint: 25 cents + 5 cents = $\frac{30}{100}$ or $\frac{3}{10}$)

3.

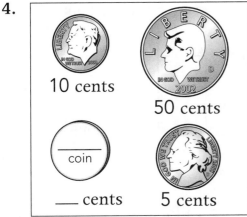

_____ cents

$\frac{1}{2}$ of 1 dollar

4.

10 cents

50 cents

_____ cents   5 cents

$\frac{3}{4}$ of 1 dollar

5.

_____ cents   50 cents

$\frac{3}{4}$ of 1 dollar

6.

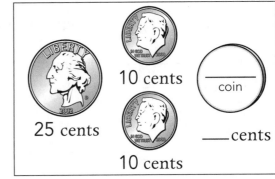

25 cents   10 cents

_____ cents

10 cents

$\frac{1}{2}$ of 1 dollar

7.

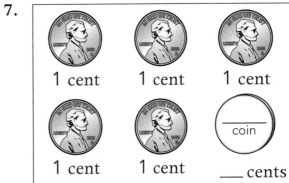

1 cent   1 cent   1 cent

1 cent   1 cent   _____ cents

$\frac{1}{10}$ of 1 dollar

8.

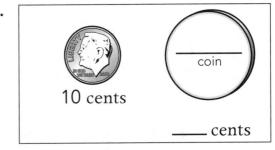

10 cents   _____ cents

$\frac{6}{10}$ of 1 dollar

Fill Up the Gallons

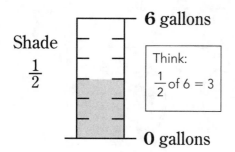

Shade $\frac{1}{2}$

6 gallons

0 gallons

Think:
$\frac{1}{2}$ of 6 = 3

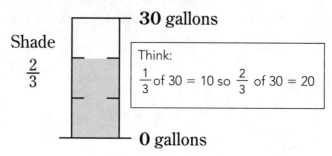

Shade $\frac{2}{3}$

30 gallons

0 gallons

Think:
$\frac{1}{3}$ of 30 = 10 so $\frac{2}{3}$ of 30 = 20

A. $\frac{1}{2}$ of 6 gallons = _____ gallons
<u>fill in</u>

B. $\frac{2}{3}$ of 30 gallons = _____ gallons
<u>fill in</u>

Shade the drawings and then fill in the blanks.

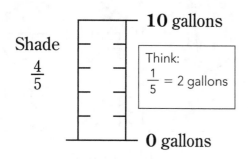

Shade $\frac{4}{5}$

10 gallons

0 gallons

Think:
$\frac{1}{5}$ = 2 gallons

1. $\frac{4}{5}$ of 10 gallons = _____ gallons

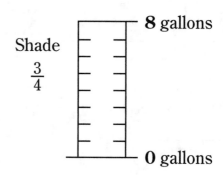

Shade $\frac{3}{4}$

8 gallons

0 gallons

3. $\frac{3}{4}$ of 8 gallons = _____ gallons

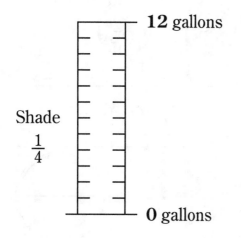

12 gallons

Shade $\frac{1}{4}$

0 gallons

2. $\frac{1}{4}$ of 12 gallons = _____ gallons

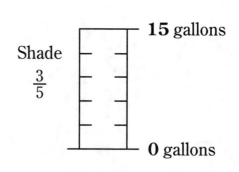

Shade $\frac{3}{5}$

15 gallons

0 gallons

4. $\frac{3}{5}$ of 15 gallons = _____ gallons

Find the Measurements

1. Show $\frac{1}{4}$ full

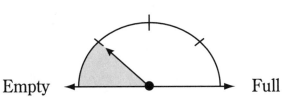

2. Show $\frac{3}{4}$ full

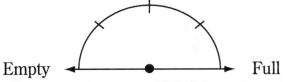

3. Show $\frac{5}{8}$ full

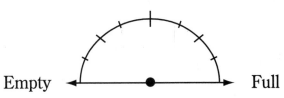

4. Show $\frac{7}{8}$ full

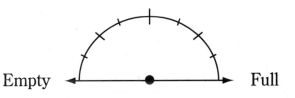

Use the pictures to answer the questions.

5. Shade $\frac{2}{3}$ full

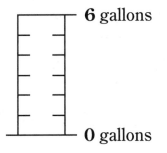

6. $\frac{1}{3}$ of 6 gallons = _____

7. $\frac{2}{3}$ of 6 gallons = _____

8. $\frac{3}{3}$ of 6 gallons = _____

9. $\frac{1}{2}$ of 6 gallons = _____

10. Shade $\frac{3}{4}$ full

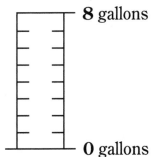

11. $\frac{1}{4}$ of 8 gallons = _____

12. $\frac{1}{2}$ of 8 gallons = _____

13. $\frac{3}{4}$ of 8 gallons = _____

14. $\frac{4}{4}$ of 8 gallons = _____

Measurement Review

Draw lines for the following lengths.

1. $3\frac{7}{8}$

2. $2\frac{1}{4}$

3. $4\frac{3}{4}$

4. $5\frac{9}{16}$

Solve each problem.

5. 9 nickels = $\dfrac{\square}{\square}$ of a dollar

7.

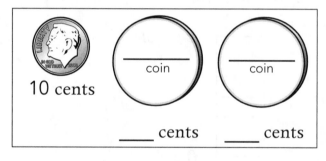

10 cents   _____ cents   _____ cents

$\frac{1}{4}$ of 1 dollar

6. Show $\frac{3}{8}$ full

Empty ⟷ Full

8. Shade $\frac{1}{4}$ full

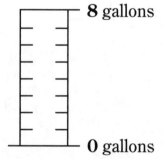

8 gallons

0 gallons

Fractions: The Meaning of Fractions

Real-Life Applications

To find one part of a total, divide the denominator into the total.

$\frac{1}{4}$ of an hour = $\frac{1}{4}$ of 60 minutes

$$
\begin{array}{r}
15 \text{ minutes} \\
4\overline{)60} \text{ minutes} \\
\underline{4} \\
20 \\
\underline{20} \\
0
\end{array}
$$

Fractions of an Hour

1 hour = 60 minutes

1. $\frac{1}{2}$ of an hour is how many minutes? _____

2. $\frac{1}{6}$ of an hour is how many minutes? _____

3. $\frac{1}{3}$ of an hour is how many minutes? _____

4. $1\frac{1}{4}$ hours is how many minutes? _____

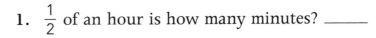

60 minutes + ? minutes

Fractions of a Year

1 year = 12 months

5. $\frac{1}{2}$ of a year (semi-annual) is how many months? _____

6. $\frac{1}{4}$ of a year (quarterly) is how many months? _____

7. $\frac{1}{3}$ of a year is how many months? _____

8. $1\frac{1}{2}$ years is how many months? _____

12 months + ? months

Fractions of a Dollar

1 dollar = 100 cents

9. $\frac{1}{2}$ of a dollar is how many cents? _____

10. $\frac{1}{4}$ of a dollar is how many cents? _____

11. $\frac{3}{4}$ of a dollar is how many cents? _____
(Hint: use your answer to question 10 as a starting point.)

12. $\frac{1}{10}$ of a dollar is how many cents? _____

Life-Skills Math Review

1. $\frac{1}{6}$ of an hour is how many minutes? _____

2. $1\frac{1}{3}$ years are how many months? _____

3. Geneva wants only $\frac{1}{3}$ as much orange punch as the recipe calls for. How much would the new recipe take?

New Recipe	Original Recipe
a) ____	9 oranges
b) ____	3 gallons of ice cream
c) ____	12 cups of ginger ale

4. What fraction of a dollar is shown? _____

quarter   dime   nickel   dime

Simplify the fractions.

5. $\frac{8}{12} =$ 6. $\frac{4}{28} =$ 7. $\frac{32}{40} =$ 8. $\frac{18}{36} =$

9. $\frac{10}{12} =$ 10. $\frac{6}{15} =$ 11. $\frac{15}{25} =$ 12. $\frac{16}{30} =$

Change each fraction to a decimal.

13. $\frac{7}{8}$ 14. $\frac{12}{15}$ 15. $\frac{5}{4}$

1. Shade $\frac{3}{8}$.

6. Rewrite as a decimal.

   $\frac{7}{8} =$

   Answer: _____

2. Compare using $<$, $>$, or $=$.

   $\frac{4}{6}$ ◯ $\frac{13}{18}$

7. Find the decimal form. Divide to the hundredths place and write the remainder as a fraction.

   $\frac{35}{20} =$

   Answer: _____

3. Find the GCF of 3 and 12.

   Answer: _____

8. What fraction of a dollar is 12 nickels?

   Answer: _____

4. Find the GCF.

   $\frac{12}{18}$

   Answer: _____

9. Shade $\frac{1}{3}$

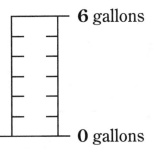

   6 gallons

   0 gallons

5. Change to lowest terms.

   $\frac{14}{20} =$

   Answer: _____

10. $\frac{1}{4}$ of an hour has how many minutes?

   Answer: _____

1. Which fraction is larger, $\frac{1}{9}$ or $\frac{1}{10}$?

Answer: _____

2. Change $\frac{1}{3}$ to an equivalent fraction with a denominator of 30.

Answer: _____

3. Change $\frac{14}{21}$ to a fraction in simplest form.

Answer: _____

4. Which two numbers are represented in this drawing: $\frac{7}{4}$, $\frac{3}{7}$, or $1\frac{3}{4}$?

Answer: _____

5. What is the decimal value of $\frac{5}{8}$ to three decimal places?

Answer: _____

6. Which is largest: $\frac{7}{8}$, $\frac{5}{8}$, or $\frac{3}{8}$?

Answer: _____

7. Write $\frac{10}{24}$ in simplest form.

Answer: _____

8. In a class of 30 registered students, 6 are absent. What fraction of the students are absent? Express the answer in simplest form.

Answer: _____

9. Change $\frac{5}{6}$ to an equivalent fraction with a denominator of 24.

Answer: _____

10. Write $\frac{13}{5}$ in decimal form.

Answer: _____

11. Phil was at bat 15 times, and he got 6 hits. What fraction of his times at bat did he get hits?

Answer: _____

12. Change $\frac{5}{8}$ to an equivalent fraction with a denominator of 40.

Answer: _____

13. If a whole is divided into 9 equal parts, then 1 of the parts is what fraction of the whole?

Answer: _____

14. Find the decimal value of $\frac{27}{20}$.

Answer: _____

15. Change $\frac{11}{12}$ to an equivalent fraction with a denominator of 48.

Answer: _____

16. Pilar has to drive 100 miles. She has already driven 65 miles. What fraction of the total distance has she driven? Express the answer in simplest form.

Answer: _____

17. Change $\frac{15}{36}$ to a fraction in simplest form.

Answer: _____

18. There are 20 families in Maria's apartment building. 16 of the families have children living at home. What fraction of the families in the building have children living at home? Express the answer in simplest form.

Answer: _____

19. Write $\frac{25}{45}$ in simplest form.

Answer: _____

20. Find the decimal form of $\frac{5}{12}$ to the hundredths place. Write any remainder as a fraction.

Answer: _____

Evaluation Chart

On the following chart, circle the number of any problem you missed. The column after the problem number tells you the pages where those problems are taught. You should review the sections for any problem you missed.

Skill Area	Posttest Problem Number	Skill Section	Review Page
Identifying Fractions	4, 13	7–19	20
Comparing Fractions	1, 6	21–25	26
Equivalent Fractions	2, 9, 12, 15	27–33	34
Comparisons	8, 11, 16, 18	35–39	40
Simplifying Fractions	3, 7, 17, 19	41–57	58
Fractions and Decimals	5, 10, 14, 20	59–62	63
Life-Skills Math	All	64–73	72, 74

denominator   the bottom part of a fraction

$$\frac{5}{8} \leftarrow$$

divisible   a number that can be divided, usually with no remainder

The number 12 can be divided by 1, 2, 3, 4, 6, and 12.

equivalent fraction   fractions that have the same value

$$\frac{1}{3} = \frac{2}{6}$$

factor   pairs of numbers that multiply to form a given number

$1 \times 12 = 12$
$2 \times 6 = 12$
$3 \times 4 = 12$
1, 2, 3, 4, 6, and 12 are all factors of 12.

fraction   a way of showing parts of a whole; made up of two numbers, the numerator and the denominator

$\frac{1}{2}$ ◄—— part; numerator
◄—— whole; denominator

gallon   a customary measurement for liquid

1 gallon = 3.8 liters

greatest common factor   the largest number that two numbers can be divided by

What is the greatest common factor of 12 and 8?

The factors of 12 are: 1×12, 2×6, 3×4.

The factors of 8 are: 1×8, 2×4.

The largest common factor is 4.

hour   a measure of time

1 hour = 60 minutes

improper fraction   a fraction where the numerator is greater than the denominator

The improper fraction $\frac{7}{5}$

can be rewritten as $1\frac{2}{7}$.

mile   a customary measurement for distance

1 mile = 1.6 kilometers

mixed number   the combination of a whole number and a fraction

$2\frac{2}{3}$ is a mixed number.

numerator   the top part of a fraction

$$\frac{5}{8} \leftarrow$$

pound   a customary measurement for weight

$$2.2 \text{ pounds} = 1 \text{ kilogram}$$

product   the answer to a multiplication problem

$$\begin{array}{r} 4 \\ \times\, 4 \\ \hline 16 \leftarrow \end{array}$$

proper fraction   a fraction where the numerator is smaller than the denominator

$\frac{5}{8}$ is a proper fraction.

remainder   the number that is left over when a division problem doesn't divide evenly

$$4 \overline{)17} \quad 4 \text{ R}1 \downarrow$$

simplify (reduce)   to make the numbers in a fraction smaller without changing the value of the fraction

$$\frac{4}{6} = \frac{2}{3}$$

sum   the answer to an addition problem

$$\begin{array}{r} 2 \\ +\, 2 \\ \hline 4 \leftarrow \end{array}$$

symbol   a written sign used to represent an operation

$$2 + 2 = 4 \qquad 5 - 3 = 2$$
$$\quad\uparrow \qquad\qquad \uparrow$$
$$4 \times 4 = 16 \qquad 20 \div 5 = 4$$
$$\quad\uparrow \qquad\qquad\quad \uparrow$$

whole number   a number beginning with 0, 1, 2, 3, and so on